Math Is Every

MATH BY THE OCEAN

By Kieran Shah

Gareth Stevens
PUBLISHING

Please visit our website, www.garethstevens.com. For a free color catalog of all our high-quality books, call toll free 1-800-542-2595 or fax 1-877-542-2596.

Library of Congress Cataloging-in-Publication Data

Names: Shah, Kieran.
Title: Math by the ocean / Kieran Shah.
Description: New York : Gareth Stevens Publishing, [2017] | Series: Math is everywhere! | Includes index.
Identifiers: LCCN 2016027684| ISBN 9781482455083 (pbk. book) | ISBN 9781482455090 (6 pack) | ISBN 9781482455106 (library bound book)
Subjects: LCSH: Arithmetic–Juvenile literature. | Seashore–Juvenile literature.
Classification: LCC QA115 .S497 2017 | DDC 513–dc23
LC record available at https://lccn.loc.gov/2016027684

First Edition

Published in 2017 by
Gareth Stevens Publishing
111 East 14th Street, Suite 349
New York, NY 10003

Copyright © 2017 Gareth Stevens Publishing

Editor: Therese Shea
Designer: Sarah Liddell

Photo credits: Cover, p. 1 Gabriele Maltinti/Shutterstock.com; p. 5 Thaiview/Shutterstock.com; pp. 7, 24 (footprint) David Herraez Calzada/Shutterstock.com; p. 9 xpixel/Shutterstock.com; p. 11 Stephen Rees/ Shutterstock.com; pp. 13, 17, 24 (starfish) EllenSmile/Shutterstock.com; pp. 15, 17 (shells) Nina Kesareva/ Shutterstock.com; pp. 19, 24 (crab) javarman/Shutterstock.com; p. 21 (top left, top middle, and bottom) Stepan Bormotov/Shutterstock.com; p. 21 (top right) Coprid/Shutterstock.com; p. 23 XiXinXing/ Shutterstock.com.

Printed in China

CPSIA compliance information: Batch #CW17GS: For further information contact Gareth Stevens, New York, New York at 1-800-542-2595.

Contents

It's a hot day.
We're going
to the ocean!

We count footprints
in the sand.
My baby brother
makes 4 footprints!

We make shapes
with sand.
I make a triangle.
My sister makes a square.

9

We write numbers
in the sand.

1 2 3 4 5

6 7 8 9 10

We find 6 starfish.
They aren't fish!
They're called
sea stars, too.

We find 5 shells.
Animals used to live
in them!

There are more starfish than shells.

There's a crab!
It has 10 legs.

I have 3 balls.
My friend has 1 ball.
We have 4 balls in all.

It's time to go.
The ocean was fun!

Words to Know

crab

footprint

starfish

Index

24